YOUR KNOWLEDGE HAS VALUE

- We will publish your bachelor's and
  master's thesis, essays and papers

- Your own eBook and book -
  sold worldwide in all relevant shops

- Earn money with each sale

Upload your text at www.GRIN.com
and publish for free

Md. Siddiqur Rahman, Salena Akter

Saline Zone at the Sundarbans Mangrove Forest of Bangladesh

GRIN Publishing

Bibliographic information published by the German National Library:

The German National Library lists this publication in the National Bibliography; detailed bibliographic data are available on the Internet at http://dnb.dnb.de .

Imprint:

Copyright © 2009 GRIN Verlag GmbH
Print and binding: Books on Demand GmbH, Norderstedt Germany
ISBN: 978-3-656-84847-9

This book at GRIN:

http://www.grin.com/en/e-book/283836/saline-zone-at-the-sundarbans-mangrove-forest-of-bangladesh

Saline Zone at the Sundarbans Mangrove Forest of Bangladesh

Md. Siddiqur Rahman[1*] and Salena Akter[2]

[1] M.S. in Forestry, Institute of Forestry and Environmental Sciences, University of

Chittagong, Chittagong-4331, Bangladesh

[2] Department of Environment, Narsingdi Office, Ministry of Environment and Forest,

Narsingdi 1600, Bangladesh

[*]Corresponding Author:

Md. Siddiqur Rahman

Address: Institute of Forestry and Environmental Sciences, University of Chittagong,

Chittagong 4331, Bangladesh

Abstract

The Sundarbans in Bangladesh part is one of the largest mangrove forests in the tropics comprising about six thousand square kilometers of forest area. It possesses unique character of having ebb and tide at every six hours, hence having salinity intrusion in its rivers and canals from Bay of Bengal. The forest is experiencing heavy saline intrusion in its lower saline zones. Therefore, there is a need to enumerate the saline water intensity in various zones of the forest. Present study tries to find out how much saline intensity exists in different rivers and canals of the Sundarbans. Results depicts that saline intensity was much higher in southwestern part, where it is moderate in northeastern part. Present study reveals the range of P^H values between 6.5 and 7.7. P^H was found to be lower in Kachikhali and Karamjol points, where Katka, Dublar char and Hiron point shows higher value i.e. near alkaline water. Western zone is higher saline intensive than eastern part. Electrical conductivity found to be between 5.2 and 24 dSm^{-1} (desi siemens per meter) in different sample collection points. Amount of NaCl (mg/L) was found between 15 and 30 mg/L. Therefore, the Sundarbans river systems and canals have strong saline intensity and the water quality seems to be brackish.

Keywords

Saline intensity, P^H and EC, NaCl, Saline zone, Bangladesh Sundarbans

Introduction

The Sundarbans is the largest single block of tidal halophytic mangrove forest in the world (Banglapedia 2006). It has been recognized as an international Ramsar Wetland Site and declared as a World Heritage Site by the UNESCO (United Nations Educational Scientific and Cultural Organization) in 1997 (Rashid et al. 2008). Total area of the World Heritage Site is 1400 sq. km. out of which 910 sq. km. is land and 490 sq. km. is water (Banglapedia 2006). Sundarbans is a region of transition between the freshwater of the rivers originating from the Ganges and the saline water of the Bay of Bengal (Wahid et al. 2002). The forest has been divided into three ecological zones, based on salinity and distribution of species composition such as i) less saline/fresh water zone, ii) moderately salt water/moderately saline zone and iii) salt water zone/strong by saline zone (Karim 1988).

Here saline water intrusion is highly seasonal. It is at its minimum during the monsoon (June-October) when the main rivers discharge about eighty percent of the annual fresh water flow. In dry season months, the saline front begins to penetrate inland, and the affected areas rise sharply from 10 percent in the monsoon to over 40 percent. Extreme weather events induced by climate change, especially low flow conditions in the dry season, will accentuate the saline intrusion in the coastal areas (Islam and Ahmed 2004).

Sundarbans might risk adding one more stress to a fragile ecosystem that will likely be critically impacted by sea level rise and salinity concerns (Agarwala et al. 2003). The SRF is presently undergoing fundamental, long-term ecological changes due to progressing sedimentation, siltation, subsiding tectonic movements, and possibly sea level changes.

The forest floor varies from 0.9 m to 2.11 m above sea level (Katebi and Habib 1987). The forest consists of numerous creeks and rivulets which play a crucial role in bringing a balance between saline and fresh water: the former being brought by semi-diurnal tides, and the latter through rivers and precipitation which helps continuously flush the salinity off the forest floor. Freshwater tends to dominate during the monsoon season while salinity levels are highest in the dry season that precedes the monsoons (Agarwala et al. 2003). Clarke and Hannon (1970) reported salinity to have the major effect on the mangrove zonation patterns and are correlated with tree height gradients. Therefore, it could be very much susceptive to heavy salinity intrusion other than its normal threshold level. Thus, a need may exist to measure and identify locations with high saline intensity in different places of Sundarbans.

Materials and methods

Materials

To identify water quality in the forest rivers and canals following instruments were used such as P^H meter, Electrical conductivity meter and Refractometer were used to measure P^H, Electrical conductivity and saline intensity of sample water respectively. Laboratory analysis was conducted to measure the amount of NaCl (gm/L) found in sample water.

Study site

The Sundarbans originally measured (about 200 years ago) to be of about 16700 km². Now it has reduced to about 1/3 of the original size (Wahid et al. 2002). The forest lies between 89° 00' and 89° 55' east longitudes and 21° 30' and 22° 30' north latitudes. It has an area of 5770 km^2, of which 4016 km^2 covered by the forests and the remaining 1756 km^2 are in the form of rivers, canals and creeks, varying from a few meters to several miles (Hussain and Acharya 1994). Forest Resources Management Project (FRMP), 1998 (modified) reported the area as 6017 km^2 where rivers occupy a major portion (1905 km^2), about 32 percent of the total area. Figure 1 indicates the location of sample collection points in the Sundarbans reserve forest and it's administrative and conservation sites.

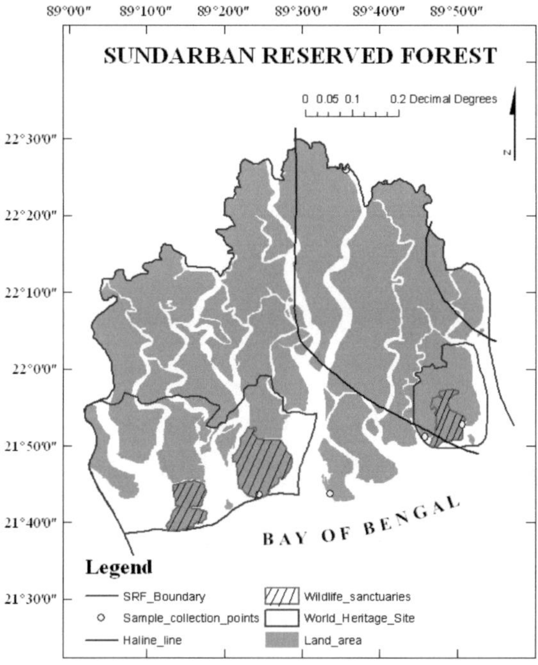

Figure 1. Location of sample collection points in the Sundarbans reserve forest (Source: GIS map by Author, Md. Siddiqur Rahman)

The Sundarbans Reserve Forest (SRF) comprises 45 percent of the productive forest of the country (Hussain and Karim 1994). SRF Constitute about 51% of the total reserved forest estate of Bangladesh (FAO 1995).

Physiography

Sundarbans is composed of three wildlife sanctuaries. Sundarbans East Wildlife Sanctuary extends over an area of 31,227 ha in freshwater zone, Sundarbans South Wildlife Sanctuary extends over an area of 36,970 ha where there is evidently the greatest seasonal variation in salinity levels and possibly represents an area of relatively longer duration of moderate salinity and Sundarbans West Wildlife Sanctuary extends over an area of 71,502 ha at the western brackish site. The three sanctuaries are intersected by a complex network of tidal waterways, mudflats and small islands of salt tolerant mangrove forest (Banglapedia 2006).

Collection of saline water sample from the Sundarbans

Saline water from the rivers and streams in the Sundarbans were collected from five different stations. For each stations, ten water samples were collected and preserved in airtight bottled where no chance of being changed in concentrations were present. The methodology followed for collecting sample water from stations was similar to that of Faruk-E-Azam and Zaman (2010) and Rashid et al. (2008). Collection of saline water was accomplished at five different stations, viz. Karamjol, Kachikhali, Katka sea beach, Dublar char and Hiron point (here the local names have been used).

Table 1 describes the respective rivers from which the sample was taken at different stations of the forest. It showed that different stations fall within different rivers and canals. Water p^H was measured according to Ghosh et al. (1983) using glass electrode p^H meter. Electrical conductivity (EC) was measured using conductivity meter according to Tandon (1995).

Table 1. General information about the water sample collection points in SRF

Stations	Forest Range	River	Saline intensity
Karamjol	Sharankhola	Passur	Low
Kachikhali	Sharankhola	Haringhata	Low
Katka sea beach	Sharankhola	Shila gang	Mid
Dublar char	Sharankhola	Passur	Mid
Hiron point	Khulna	Arpangasia	High

Procedure to determine the amount of salt present in sample water

Determination of Sodium Chloride (NaCl) in seawater according to Mortimer (1941) was used to measure the amount of common salt present in seawater, mud and other ingredient. Following equations represent the measurement procedure for NaCl in marine water.

$$\text{Amount of NaCl} = \frac{15 - \text{ml } 0.1\text{N titer} \times 0.585}{\text{weight of sample}} \qquad (1)$$

$$= \frac{(15 - \text{ml } 0.1\,\text{N NH}_4\text{SCN} \times 0.585)\,\text{gm}}{\text{weight of sample}} \qquad (2)$$

Results

Three different salinity parameters were taken into consideration for measuring the saline conditions of the rivers and canals of the Sundarbans reserve forest. Table 2 shows the results found for different parameters in five different stations.

Table 2. Saline intensity, P^H value, Electrical conductivity and amount of NaCl (g/lt) found in sample water

Stations	Saline intensity ppt (S°/00)	P^H	E_C Desi siemens /cm	NaCl (mg/L)
Karamjol	10	6.7	15	15.07
Kachikhali	3	6.5	5.2	16.56
Katka	12	7.5	16	29.47
Dublar char	20	7.7	24	25.40
Hiron point	18	7.5	23	21.10

Saline intensity and electrical conductivity of river water was found higher at Dublar char and Hiron point stations where lower at Kachikhali and slightly moderate at Karamjol and Katka. Water P^H increases from eastern part to western part indicating presence of high base captions at those sites especially concentrated Na$^+$ and Mg^{2+}. Table 2 indicates amount of NaCl was higher in Katka sea beach followed by Dublar char and Hiron point both of which are sea side mangroves, where low concentration was found at Karamjal and Kachikhali point of being offshore mangroves. For this, whole mangrove forest might be divided into a line transact at the centre dividing it through north-east, north-west, south-east and south-west as indicated in Figure 2. South and western part of the forest experiences higher saline condition where northern and eastern part has low saline intensity in river and canal waters.

Figure 2. Salinity zonation in general for the Sundarbans Reserved Forest (SRF) showing different compartments of the forest (Source: GIS map by Author, Md. Siddiqur Rahman)

Discussion

P^H for all the rivers in the Sundarbans was found to be between 7.1 and 7.5 (Faruk-E-Azam and Zaman 2010) and between 7.5 and 7.9 in four different stations viz. Karamjal, Kochikhali, Kotka and Dublar Char (Rashid et al. 2008). Where, present study reveals the range between 6.5 and 7.7. Thus the water samples do not changed significantly with respect to P^H value in the study area (Table 2). However, the range widens from previous results found. There might be wide fluctuations in P^H at those points. P^H was found to be lower in Kachikhali and Karamjol points, where Katka, Dublar char and Hiron point shows higher P^H value that means near alkaline water. This reveals high saline intensity in later three points. Moreover, first two points are located in eastern part while later three are on western part of the forest. Thus, western zone might be higher saline intensive than eastern part.

Electrical conductivity to be between 5.2 and 24 dSm^{-1} (desi siemens per meter) in different sample collection points, where of all the rivers, EC was found to be between 24.8 and 34.0 dSm^{-1} (Faruk-E-Azam and Zaman 2010). This might be due to seasonal change in river and canal water in those sites.

Amount of NaCl (mg/L) was found to be between 15 and 30 mg/L in different site points of the forest. This amount denotes that the rivers and creeks are truly saline prone. Meade (1989) found that among the dominant cations, higher concentration of Na caused toxicity (acceptable range < 75.0 mgL^{-1}) in water of Sundarbans. Concentration of sodium chloride found to be imposing salinity in the rivulets f the forest jurisdiction. According to Figure 2, the forest area was divided into four different salinity zones such as very low, moderate low, saline and strongly saline. This sort of deviation was conducted for results found for different saline points and other results.

The soil of the Sundarban is saline due to tidal interactions, although the salinity is low compared to soil salinity in other mangrove forests of the world (Karim 1988). Soil salinity, however, is regulated by a number of other factors including surface runoff and groundwater seepage from adjacent areas, amount and seasonality of rainfall, evaporation, groundwater recharge and depth of impervious subsoil, soil type and topography etc. It is found that, conductivity of subsurface soil is much higher than that of surface soil. Using the salinity scale, the forest areas have been divided into three zones based on soil salinity (Karim 1988).

Figure 3 shows how saline intensity in ppt shift from east to west part of the forest. Sundarbans might risk adding one more stress to a fragile ecosystem that will likely be critically impacted by sea level rise and salinity concerns. The effect of water diversion upstream on dry season flows and salinity levels in the Sundarbans was in fact comparable to (if not higher than) the impact that might be experienced several decades later as a result of climate change. Since the 1980s coastal lands have also been extensively brought under shrimp cultivation – primarily in response to the high salinity. Freshwater tends to dominate during the monsoon season while salinity levels are highest in the dry season that precedes the monsoons (Agarwala et al. 2003).

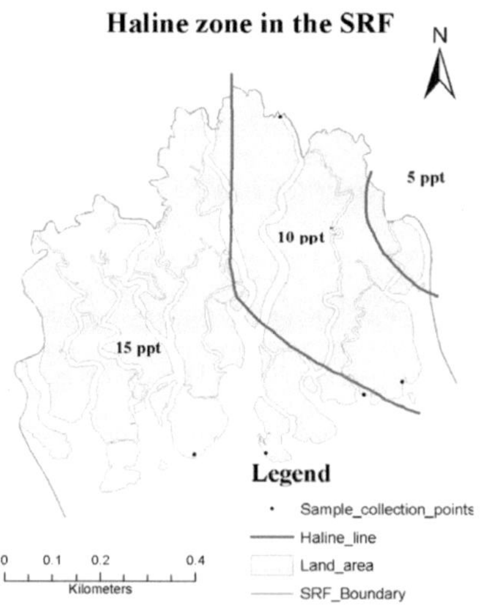

Figure 3. Zonation of salinity in the Sundarbans. Adapted from Karim (1988)

According to salinity, there are three zones in the Sundarbans viz. oligohaline, mesohaline and poloyhaline zones (Curtis 1933; Karim 1994; Siddiqi 2001). Based on soil salinity distribution, three distinct salinity zones - oligohaline (salinity<2 dsm^{-1}), mesohaline (salinity 2–4 dsm^{-1}) and polyhaline (>4 dsm^{-1}) zones could be recognized (Siddiqi 2001).

Iftekhar and Saenger (2008) identified location, impact zone and salinity zones of the Sundarbans Reserved Forest in Bangladesh in three categories. From Iftekhar and Saenger's (2008) GIS map (Figure 4), there found shift in saline intensity from east to west and also from north to south.

Figure 4. Three salinity zone in the Sundarbans reserved forest (Iftekhar & Saenger 2008)

Conclusion

Sundarbans is the largest single track of tidal mangrove forest in the world, located in the Southwestern part of Bangladesh. It lies on the Ganges-Brahmaputra delta at the point where it merges with the Bay of Bengal. In Sundarbans, east side contains moderate to low saline water in river and streams but in western side, here contains moderate to high saline intensity in rivers and streams. On the other hand, salinity increases from North to south because at northern part where fresh water from upstream is high and sea (Bay of Bengal) water is low.

Acknowledgement

Authors would like to thank the tour management committee of the Institute of Forestry and Environmental Sciences, University of Chittagong, 2009 for their support during the study.

References

Agarwala S, Ota T, Ahmed AU, Smooth J, Aalst MV. 2003. Development and Climate Change Bangladesh: Focus on Coastal Flooding and the Sundarbans, Organization for Economic Co-operation and Development (OECD), Paris.

Banglapedia. 2006. National Encyclopedia of Bangladesh. Asiatic Society of Bangladesh. Dhaka. http://www.banglapedia.org/httpdocs/HT/S_0602.HTM. Accessed: 25 Jan 2012

Clarke LD, Hannon NJ. 1970. The mangrove swamp and salt marsh communities of the Sydney district. III. Plant growth in relation to salinity and waterlogging. Journal of Ecology. 58: 351-369.

Curtis SJ. 1933. Working plan for the forest of the Sundarbans division for the period 1931-57, Vol. 2 and appendix. Bengal Government Press, Calcutta.

FAO. 1995. Integrated Resource Management Plan of the Sundarbans Reserved Forest – Final Report. FAO Project BGD/84/056. Food and Agricultural Organization of the United Nations, Rome, Italy.

Faruk-e-Azam AKM, Zaman MW. 2010. Water Quality Assessment After Sudden Change of Color at the Sundarban Mangrove Forest. SAARC Journal of Agriculture. 8 (2): 76-86.

Ghosh AB, Bajaj JC, Hassan R, Singh D. 1983. Soil and Water Testing Methods. Laboratory Manual, Division of Soil Sci. Agric. Chem., IARI, New Delhi, India, pp 48.

Hussain Z, Acharya G. 1994. Mangroves of the Sundarbans. Volume two: Bangladesh. IUCN, Bangkok, Thailand, pp 256.

Hussain Z, Karim A. 1994. Introduction. In: Hussain, Z. and G. Acharya (eds.). Mangrove of the Sundarbans, Volume 2: pp. 1-10. IUCN-The World Conservation Union, Bankok, Thailand.

Iftekhar MS, Saenger P. 2008. Vegetation dynamics in the Bangladesh Sundarbans mangroves: a review of forest inventories. Wetlands Ecological Management. 16: 291–312.

Islam MR, Ahmed M. 2004. Living in the Coast: Problems Opportunities and Challenges. PDO-ICZMP, Water Resources Planning Organization (WARPO), Dhaka, Bangladesh.

Karim A. 1994. Physical environments. In: Mangroves of the Sundarbans. Volume two: Bangladesh. Edited by HUSSAIN, Z. & G. ACHARYA (1994). IUCN, Bangkok, Thailand, pp 11-42.

Karim A. 1988. Environmental factors and the distribution of mangroves in Sundarban with special reference to Heritiera fomes. Buch.-Ham. PhD. Thesis, University of Calcutta, pp 222.

Katebi MNA, Habib MG. 1987. Sundarbans and Forestry in Coastal Area Resource Development and Management Part II. BRAC Printers, Dhaka, Bangladesh pp 107.

Meade JW. 1989. Aquaculture management. Van Nostrand Reinhold, New York.

Mortimer CH. 1941. The Exchange of Dissolved Substances Between Mud and Water in Ocean. Journal of Ecology. 29: 280-329.

Rashid SH, Bocker R, Hossain ABME, Khan SA. 2008. Undergrowth species diversity of Sundarban mangrove forest (Bangladesh) in relation to salinity. Ber. Inst. Landschafts- Pflanzenökologie Univ. Hohenheim Heft 17, 2007, S. 41-56, Stuttgart 2008. available at https://ecology.uni-hohenheim.de/fileadmin/einrichtungen/ecology/Dateien_Inst Ber_17/1_NEU_OrRashid_41-57.pdf

Siddiqi NA. 2001. Mangrove forestry in Bangladesh. Institute of Forestry and Environmental Sciences, University of Chittagong, Chittagong, Bangladesh, pp 201.

Tandon HLS. 1995. Methods of Analysis of Soils, plants, water and fertilizer. New Delhi: Fertilizer. Development of Consultation Organization, India.

Wahid SM, Alam MJ, Rahman A. 2002. "Mathematical River modeling to support ecological monitoring of the largest mangrove forest of the world – the Sundarbans". Proceedings of First Asia-Pacific DHI software conference, 17–18 June, 2002.